I0469771

THE HAND OF GOD IN THE TIME TRAVELER'S UNIVERSE

Reasons To Believe

By
Daniel W. Merrick PHD

ISBN-13: 978-1519100573

ISBN-10: 1519100574

The Hand Of God In The Time Traveler's Universe; Copyright
©1997-2015 Daniel W. Merrick PHD, Eternal Light & Power
Co. Publishing. All Rights Reserved. www.Yahbible.com

FORWARD

I originally wrote this in 1997 and posted it online as proof of reasons to believe in God and a Creator as opposed to the prevailing evolutionary religion. I call it a religion because of the content of this book and the theory with which I have proven that evolution is false within the pages of this book.

I have added updates to the reasons with current DNA and Genealogical science that prove further causes to have a faith in the bible as opposed to faith in evolution.

I hope you will enjoy this paper and share this book with those who may doubt that God is and by these words find a new faith in the creator that loves us all so very much.

Table of Contents

The Beginning

Genesis 1:1 In the beginning God created

" What ever the mind of man can conceive and believe, it can achieve " Zig Ziglar

Although the origin of the above statement is most likely J.P. Morgan, Dale Carnegie, or most likely predates it's original author of record and copyright, it was first said somewhere and expressed in the actions of history by men like the Wright Brothers, Albert Einstein, and somewhere tonight, in the mind of a child who will change life as we know it forever. I know this is true because it is my story and yours. It is the child laughed at for having a crazy idea. It is the person rejected for being different. It is the science of amateurs who find the greatest discoveries because they see things differently than the experts. We all know these people. We do all we can to keep them from getting our jobs. We do all we can to make them look bad. Even when our intentions are subconscious, our actions are to isolate these genus creatures from society. When I was a boy they called them nerds. Today all the world looks to them for the power to take the small minded, superficial, and the plastic into time itself and come back a new creation.

One of my geek pet studies is physics. I can't get enough. I read Stephen Hawking, Uncle Albert, as I like to call the big E-man, (as in $E=MC^2$ [squared]), and when the Discover Channel has one of those 'Awe' programs on, I force my kids to watch it and we talk about small town thinking in a big city world. Some years back I had a friend that worked at NASA who would talk with me about the nature of the universe and what the future would be like. I had a theory about magnetism that I shared with him and his friend one Saturday afternoon. That idea became the spark of thought that lead to the physics behind the digital tuner. I did not get a penny for my contribution to the though process. But who cares, my life is better for it.

Now I am going to do it again. And it will be little wonder to me if someone else has thought of it already, in that I have talked to many about this idea already. Besides, great minds think alike. That's is why there is so many law suits over music and song rights. It is my belief that this is because we all have the power to tap into the creative energy of the universe. I call it the HOLY SPIRIT, and well it is, but even more from a theological stand point it was foretold to be so: "In the last days I shall pour out my Spirit upon all flesh" Joel 2 The Holy Bible. Besides, an idea un-shared in technology and science is an invitation for someone else to be the first to share it with the world. Once an idea is shared, it is no longer yours or mine, but it becomes ours. We all gain form it's gifts. This is the guiding principle of my faith and my life, to own the world unselfishly and eat the banquet of life savoring it every moment.

The problem here is to go beyond the light barrier. Yes there is a light barrier. Man has broken the sound barrier, the bounds of earth, and now we shall break the light barrier. To travel faster that the speed of light. My college physics teacher said "It can't be done, the energy of the universe is converted at the speed of light". Well, if you burn energy at the speed of light to travel at the speed of light then you are right, you can't! BUT I CAN! Because I will not burn energy to do it. I will ride energy.

Now many questions arise with this idea. What will happen if man travels at the speed of light. Will we undo the fabric of creation? Will we displace time it self? Will we alter our future by some how damaging our past? NO ! Only God can do that. He will not let us do that. But He will let us move at the speed of light. Because he gave us the dream to. "A man without vision shall parish". Then what is on the other side of the light barrier? My theory is Time in infinity or Quantum time.

We all know that one earth year is 365 days. One year on Pluto is 247 earth years. The theory of relativity prove that time is relative to distance from gravitational force. But I believe that this is not clarified enough. This gravity I like to call the gravity of energy. This gravity is directly related to the electromagnetic fields of energy in the universe. Each planet, body, field, and energy source giving connected strings of energy that ties earth to the sun, the sun to the stars, and black holes to the birth of a star on the other side of the universe and the expanding of all that is. If energy is equal to the mass of matter then matter transferred, converted, and displaced becomes energy to make matter in the universe. This energy transfer is constant and creates the filed of magnetic pull and balance that I like to call the E-plain (Energy string plain). The E-plain is the energy that exists from the conversion of sub-particles of matter into energy and back into matter again.

The E-plain has three properties that must be understood to harness it's force: 1) It has a magnetic field. 2) It has a infinite gravity. 3) It has a has a warp outside the universe. So if these three are true, then it has to be within matter and energy as a unification of all forces like the earth has a magma of elements that create magnetic fields. The E-plain is inside of the space contained in the orbit of strings.

Space tells matter how to move and matter tells space how to warp or bend around matter. The E-plain has a force in proportion to the balance of matter it warps toward. Let me explain. Matter causes space to warp around it, or to warp or bend away from matter. The E-plain warps in an opposite arch. It's gravity is drawn within matter, or energy. The relation to the matter and energy is the infinite gravity of the E-plain. Thus we can calculate the energy plain's force by measuring the gravity of the matter effected by it. We must also calculate its energy field in electromagnetic measurement. Energy forces may counteract with the E-plain or contribute to it's energy at certain points in the universe within matter or energy to maintain balance. We read it in space with radio telescopes and can plot its course and force on space and matter traveling through it. All matter travels in the effect of the E-plain. Only anti- matter can disrupt it. The greatest thing about the E-plain is that it forces time forward and outward. From creation to now it has moved the past out to the edge of the universe. The universe is expanding and so is time. It expands equal to the mass of the E-plain's force. Now we can measure the E-plain effect by the and force of magnetic flux density. We can also measure the force of flux by calculating the magnetic movement of a given particles orbital motion at a given point in space. The resulting value and frequency is the E-plain's value at that point in space. This super-string can then be plotted by flux and frequency of modulation. Thereby finding its path across a distance in space.

Each particle of matter displays it's nature. Energy also is the transferred matter in an expression of excited matter that moves with forces we can measure and see. When an atom is split we see this energy in gamma rays, X-rays, light and energy in heat, sound, and like wise in fusion we see absorbed light and energy released. Even when stars explode or implode, we have a measurement of energy and a sound in radio signals that are given off from the matter in a new state. I believe this state is where the space within matter has also imploded and the remnant left is a pulse of the original unity of forces combined within as in a pulsar.

Now pulsars give off energy like a beacon in the universe. We can read the frequency of the energy as a radio signal and measure its value. The E-plain also has a frequency which is constant in strings of energy equal to the relation of the plain to gravity, matter, and energy surrounding it. In other words, it is like a path we can trace throughout the universe and use as a map to travel by. It maps space between matter at a given point by the electromagnetic energy around that space. It is essential to understand to travel through space at the speed of light. Even more important to plot and know if you are going to travel at the speed of light. A piece of matter the size of a pin may be very dangerous to a craft at the speed of light. So we must be able to read matters signal at this subatomic level before we venture into the light barrier. It may be that the E-plain has a signal in the electromagnetic spectrum it is bouncing off matter itself. Being an amateur in cosmology, I could not know if those forces have been observed. More important I believed that this E-plain has shaped time and the universe within itself and is causing the expansion of the universe from inside this space inside matter and energy.

Matter itself, and it's energy, in this way becomes the marker to read as a form of radar to steer our spacecraft through the vastness of space at the speed of light.

The E-plain is a string of energy force measured in magnetic field density, at a given frequency or frequencies along a plotted distance in space.

Since energy is converted at the speed of light or less, we can not burn energy to propel us. For one thing it will make our navigation impossible and effect our energy sensors that read the E-plain. Present theory tends to show that the E-plain has equal values along different paths and at different points in the universe. I suspect that it most likely curves in relation to the edge and spin of the universe. It may be possible to use antimatter but only at a value different that the value and frequency of the string of energy we read to navigate by. Matter is very dense at the infinite point of rotation of the E-plain and has a rotation effect on the universe. While the planets spin so do galaxies, thus a galaxy has it's time as does the earth. The solar system spins and has it's time also. Thus so does the universe. The universes time is infinite because it is spinning on the axis of creation and expanding into infinite time and space. Still cosmology has this problem with gravity. Cosmologists say Gravity is a small force, but I say it is the greatest and the unification for which you seek.
Rather than deal with this problem it might be better to use the energy of the universe itself to propel us. It is simple and so obvious that the experts dismiss it as a crazy idea. That is why it works. Like the quote I began this paper with, here is my own: "What ever the experts criticize, utilize".

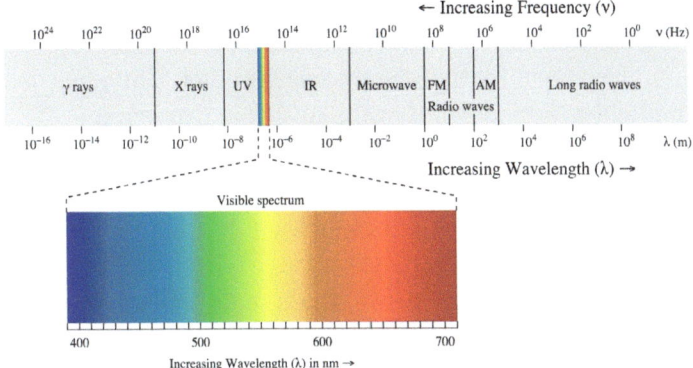

In the spectrum above what is missing? In fact, what is missing is the answer to the unified theory and math that has been right under your nose for as long as man has been searching for this answer. Einstein's Grand Unified Theory. The missing element of the chart above is sound. The spectrum of sound and vibrations. Matter is not standing still and is in movement and by that movement creates a space inside matter where strings play jumprope with matter. It is this sound waves and particles of energy that tune in the expression of matter and energy to express themselves in light, sound, and radiation. Sound inside Radio waves can create heat in a microwave and pressure in sound with large systems such as the military Active defender system that makes skin burn within cells without the effect of cell damage caused by fire. This very idea is what John Kanzius used to invent radio wave cancer therapy by bouncing waves against each other. The sound wave inside this inner-space is what vibrates strings which form all things in the universe.

By tracking and mapping energies in force and values in the universe and on earth we can enter the 21st century as the first generation that made space warp travel possible.

So you ask me 'how can this be done'? The answer is we are already doing it. When I originally wrote this book there were no patents for warp speed travel. The idea came when discovered that the universe was in fact expanding faster than the speed of light. In fact moving faster with each parsec or million light years of distance as it increased at about 35 MPH for each million light years in distance. A patent has just recently been filed for a warp speed space craft. If it works which math theory predicts is will, then mankind may be able to go around space to get to the other side of the universe in the same way light goes around the sun in a warp which we see in the test of relativity during a solar eclipse.

We know that there are 4 dimensional aspects to the universe. They are Time, Space, Matter, and Energy. Super strings are said to vibrate within 11 dimensions. But now that we have had a glimpse of antimatter, we should perhaps change our view of the dimensions to fit our model. We know of our 3D view of the universe being in geometry length, width, and depth. We to often forget volume. Density itself speaks to us of volume. So I have redefined my thinking to include length, width, and volume. Now volume has mass and area. The quantity of Space it self has it's volume in mass and area. So space, as we know it has 4 dimensions: Space Length- the distance in a straight line across the universe. Space Width- the distance in a straight line intersect space-length in the exact center, across the universe. Space Mass- the proportion of the total of space in the universe. Space Area- the value of the total surface of space at the point of expansion. Yet we have one more space unseen and blocked by matter, that being the inner orbit of space within strings. That is where the unity is.

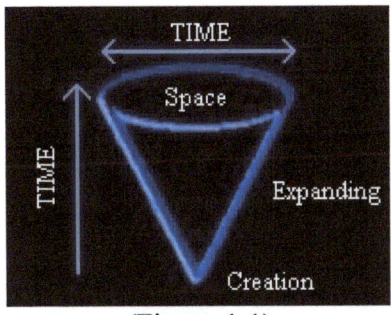

(Figure 1:1)

The cone shaped universe. In my theory of a cone shaped
universe we see time as it really is. Moving forward and
expanding with the universe it self. This is an already
proven theory and many tests have been done to prove it.
Thank you Uncle Albert. But what is the force that is
keeping this from slowing? What are we missing? We
might say that centrifugal force is making the universe
spin and maintaining the inertia to keep it moving.
Perhaps the force of the big bang has placed it all in
motion and now is slowing and it all will collapse onto
itself some day. This is what some cosmologists would
have us think. I on the other hand believe in God. My God
Yahuah would not take his hand to expand and create to
only let it spin down into another big bang and do it all
over again. That is a very crunch! pessimistic view of
things. What forces act upon space and time to make it
expand? What forces cause matter to be attracted away
from matter, galaxies away from galaxies, and hold
planets in orbit. The answer is found in Stars and black
holes and it has been suggested dark matter which is as yet
undiscovered.

Stars are the life force in the universe and when a star
dies, if it has the right constant of energy (see

Chandrasekhar limit) it becomes a black hole. Not proven to exist until now. The gravity of the universe that keeps it spinning and expanding is the gravity force of black holes.

(Figure 1:2)

The black hole catches all light, matter, energy, and time and sucks it into the vortex. The star is falling into infinite time and space toward creation in this example but it could also be a circle around the cone of the universe and act as a gravitational force at the edge of the expanding universe to draw the universe into further expansion. The existence of this gravity is not observed due to the force of its fall drawing it into the dimension outside the known universe. That is why it is in infinite force, it is on God-time in eternity. That is why I call it the Hand of God. It pushes the universe into expansion and forces it away from creation to eternity. That is why the universe will not shrink back into the universe, falling into another big bang. It reflects it's creator, The Eternal God. No big crush.

The figure above (1:2) shows that such a cone would create its own gravity that would maintain it's own spin constantly at a distance from the known universe. Thus maintaining balance between entrance to the vortex and the vast gravity it creates that is so powerful that it even captures light. The force of this gravity always moves away from the universe through infinity at an equal distance maintaining the energy constant to keep the universe spinning. If this force is toward creation or going back in time as cosmology would suggest, then time travel would be possible. If not, then this effect of gravity on the universe would maintain the spin and expansion rather like a pinwheel with the universe in the center of an elliptical shape. In such case the universe would look as such: Figure 1:3

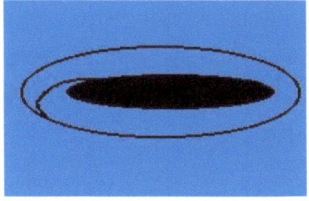

(figure 1:3)

A Black Holes fall into the E-plain (Eternal Energy Plain) and the resulting gravity spins the universe into eternal expansion. The E-space inside matter attracts it to expand.

In either case the possibility remains that travel at the speed of light may launch our traveler in the past or the future. In this universe antimatter can only exist in another dimension outside the E-plain since the antimatter would destroy the matter's fall into eternity. Thus the Black hole is a bottomless pit that forces gravity out and up as it is reflected in particle dynamics.

Time in this universe would have four dimensions and travel in all directions: Time directional =past to future
Time anti-directional =future to past
Time warp = expanding time
God Time = infinity outside the known universe

13

If this is the case then the age of the universe would have to be reduced by an exponent factor and adjusted to compensate for the expansion in all directions. Which is what we observe when we look through a telescope at the known universe. The star 100 million light years away would then appear to be much older than it is. And all the text books would have to then account for a creator. I would estimate based on my biblical studies that the universe is about 6000 years old. I will leave this formula to the cosmologists but a simple triangulation formula must also take into account the power of expansion of time which Relativity has already proven. The age of matter expands with the universe, the further away the older matter appears. Matter reflects the total time of expansion.

The Law of Time-Matter warp

$Et=MtC^2$ Energy time=Matter time Constant at the speed of light squared.

$pt-->Mt$ Radiation density effects Matter time.

$t=MC^2p$ Time = Matter Constant at the speed of light squared.

$Mp=(E=MC^2)t$ Matter radiation density=the energy& matter that is constant in the universe at the speed of light squared times time.

$ts=Mpt$ Time change=Matter radiation density time.

$ps=Ept$ Radiation density change= Energy radiation density time.

$pts^2=(Mpt)(Ept)C^2$ The above two statements are constant in the universe.

$Rs= ts^2 (K)$ The universe change=Time change squared, times the geometry of the universe which is expanding.

The time of the universe is expanding relevant to a positive expansion of it's geometry. In vastly expanding space '->' effects or results in expanded time. Therefore age is relative to size at measurement. A part of the universe can not be older than the whole, but any one part can be as old as the cumulative whole. If this were not so then the year on Pluto would be 365 days and thus on every planet in the solar system. The rotation of gravity would be to fast and there could have been no life on the Earth. It is much like the human body. Even though if one is 38 years old the cell in the body that was just born is only seconds old. When we look at the man we do not state that he is just born due to the age of his cells. The body tissue is young yet the body is old so the tissue reflects the age of the whole body even though it is only seconds old. The super-hot radiation and energy at the point of creation had just that effect on the universe. As it expanded the light that was once 2 feet from the matter of the earth increased distance and light expanded and shifted as it moved away. Determining these factors and using the basic formula r t = d where r=rate and t=time and d=distance, we can then calculate the size of the universe and check it against the rate of expansion widely accepted today of about 55km Per sec/million light years. Which is 34miles Per sec/million light years. A Light year is 186,000,000 miles times 31,536,000 seconds (the number of seconds in a year). This is based on the time it takes light to travel in space in one year. Our Galaxy, The Milky Way is 200,000 light years across in diameter and 100,000 light years wide at it's center. It takes 200,000 light years to make up one rotation around it's orbit or our galaxy has a space year of 200,000 years in the same way Pluto has a 247 year year. Now the problem with measurements until now in this science is that they keep complicating it by adding factors that are mentally predisposed to allow for the absence of God.

So they come up with figures that can't work and theory that won't wash the test of the requirement Einstein's theory has that the universe was created. Another problem is that they do not account for the expansion of the universe from it's center, they always account for it from end to end. The value needs to be half and then the expanding time factor must also be taken into account. From the center of the Milky Way Galaxy it's average radius is 31,830.914 light years. Accounting for the distance per second mega-parsec and the light year of 186 million miles we get a number of 5,920,550.004 ; Now if this number is divided by 34 (for the miles per second of expansion, our rate) we should get a number pretty close to 200,000 light years. The answer is 174,133.83 light years across the Milky Way. Adjusting for error and arc at .1877 our galaxy size would be 206,695.87 light years across. We could verify this by triangulation from the same point in space a year apart to get the exact number and rate of expansion. Now if the universe is as old as the evolution religion wants us to have faith in then, when we multiply our 34 mile rate by time of 3999999000000000000000000000000000000 or what ever outrageous number they choose to prove God wrong, then we would get a enormous number as the age of the universe and a size even beyond there present estimation. Interestingly enough, in order for creation to be true, the Bible age of about 6000 years since creation only need fit the Milky Way. And by some fate when 6000 years multiplied by 34 miles of expansion is considered with all the factors we get 204,000 light years across the Milky way. We are therefore 6000 years +or- 4 years Earth time from creation according to Biblical dating and mathematics calculation. Uniformly it puts the size of the universe at 204 billion light years and expanding into eternity. Radiation and carbon dating will prove this wrong but as the formulas above account for, Matter age is relative to the effect of the energy of the whole universe. Therefore these dating methods will always come up with an excessive amount not factoring in the time expansion effect of relativity.

Thus I have concluded that the universe is effected by expanding time and energy is reflecting it in matter time.

Some suggest the universe may have at least 12 dimensions and possibly 24. Each dimension having its own direction of time in the infinite expanse. To maintain the balance though it would most likely have 12 in the same direction of travel as we are going in time and 12 in the opposite direction. It is possible that the direction of time curves into the E-plain which would give expanse on the outer-limits of the other side to the other 12 dimensions traveling to the creation. Most likely if matter is effected toward creation then time maybe also. As the universe expands it's matter expands into a new gravitation of force which is another dimension somewhere in infinity. It's gravity having the opposite effect and pushing the universe we know and love into further expansion. More likely the reverse effect is accounted for in the E-plain itself and there are no 'Big Crunch' universes out there. Given that the rate of expansion is constant, and that the C-filed theory of gravity has been proven an error, Then the only way to account for this is Black holes having an E-plain or sphere, string, dimension on the other side that is creating a space-gravity we observe as expansion.

We can see this gravity reflected in that gravity holds us to matter and matter attracts us by gravity. The dense matter of a Black hole would therefore draw the universe in it's direction and keep us spinning happy campers forever. Without a creator this Idea would never work because something infinite would have to move this universe into motion. That's why I say 'as sure as gravity holds you to the earth, Black holes draw us toward God'. If gravity is holding us, and it is, then it is obvious to me that gravity or energy is moving us and expanding the universe. With each progression of time in figure 1:1, we have a different universe in size depending on the time from creation you observe it. A new ring of expanse can be then observed in time past and the number of black holes needed to continue expansion can be calculated. We can then know today how many black holes there are in the universe by measuring the universe's expansion and calculating the force needed to continue expansion. As new stars reach the end of there life we can then know if the creators calculations are correct. What I mean is that if there are enough stars of the right energy supply to continue expansion in the future or at the point in space-time the force is needed to continue expansion. Also if the number of new stars is equal in the equation. According to relativity it is.

The E-plain has a gravity force that is in the direction of matter and energy in the same way gravity forces us toward matter on earth. The balance of gravity between planets moves space into expansion and matter deeper into space and planets further apart. As for the cooling problem is seems to me that our sun (star) has kept us at the right temperature. And it won't burn out until well after we have moved on to other planets.

Another possibility is that the Big Bang created more than

one universe. Each having its own direction of spin and all the cones of their expanse being connected at creation. If this were so then one would have to travel at the speed of light to reach the other universe and only a worm hole or black hole could take you there. I suspect that there would have to be at least 24 of these universes to maintain symmetry and balance with the gravitational effects required to keep each expanding away from creation. This multi- dimensional parallel universes would be connected at the E-plain by black holes. Each universe would use the E-plain gravity as a jump of sorts to begin expansion at the big bang and push against each other by creating the E-plain with mutually collapsed stars for balance.

If the calculations of expansion proved that there is not a sufficient number of black holes in our universe to continue expansion then the big crunch would be the only alternative to the multi- dimensional universe theory. Without the big crunch these universes would need other universes to provide energy and gravity to continue expansion.

With either case, the E-plain theory of a single universe, 12, or 24, the energy would still remain constant with matter according to Einstein's theory. The only difference here is the Energy in the E-plain is converted into gravity unifying the process of the life of energy and matter in the universe. With multi-universes we would need at bear minimum 3 to maintain the balance of energy between the expanse of the universes.

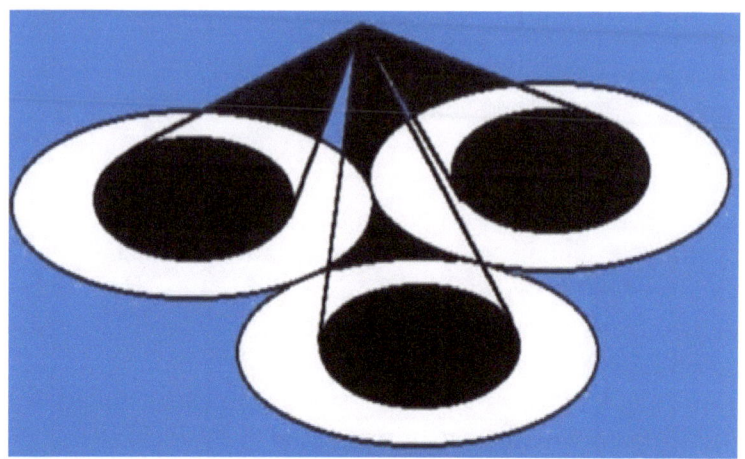

This I call the Mickey Mouse model of Cosmology. The E-plains interconnect and each universe shares it's Black Holes attraction to maintain sufficient gravitational force to continue expansion. These universes will never meet because the E-plain keeps them an equal distance in infinity from each other. As a black hole comes to an end, as with new born star's using the energy and substance of a supernova, an explosion may occur that will propel it's energy into the E-plain. This could only occur to maintain the balance of energy and matter in the universe to further expansion.

Now the problem arises that as the universe expands that it cools. An infinite expansion would therefore cause an infinite cooling. Here is where radiation has it's task. It has been discovered that the temperature of the universe has a value of 3 degrees Calvin. That this background of temperature has a frequency that maintains this constant. And one fine day in 1965 scientists found this background noise in the universe. Interesting that this sound and temperature would take on the numerical equivalent of the Trinity.

Now if the case is that this is from the big bang, which it may well be, then that minimum temperature would have well cooled in the model that cosmologists would like us to believe. Some of the universe would have to had become Zero degrees Calvin and the matter would have stopped moving. But it is not, it is expanding. So some force is maintaining that temperature throughout the vastness of space. That is the E-plain. It is God's hand keeping the universe together and preventing matter and life from coming to a complete stand still.

Now let me say that their is something about most scientists that preach evolution that I find dishonorable. That is they teach the big bang, which must have a force behind it to work, Then they escape into the world of non-definitive conjecture and start telling us about evolution of genetic material to form life. If one piece of genetic material naturally occurring in the universe that transmutation into another code of gene structure had ever occurred on this planet, we would still observe it somewhere in the chain of ecology. But we don't. Because there is none. How a man can see the vast balance of space, time, matter, and energy and still not awe at it with the clear knowledge of its clockwork like construction, is beyond me. I've heard it said that it is much like saying that an explosion occurred and in a book factory and a dictionary of every language in the world was miraculously constructed. That is beyond reason. Only with and infinite force can an infinite universe be brought into motion. This is the Law of cause and effect that few cosmologists desire to break in theories of particle physics, and continually broken in their explanation of life. Life is a unique occurrence in the universe and yet to be proven in existence anywhere other than Earth. Even

21

the Mars splash on the news a few months back proved to be 'Jumping the gun' as later reported that the scientists were mistaken. 'OOPS' one headline read in the newspaper.

Inertia is to weak to continue this expansion with out a mega-gravity force pulling the universe outward. When it comes down to it that is what most cosmologists are depending on. That inertia will slow , the expansion will slow, and that the big crunch will land on all existence in a few Billion Trillion to some power years. With that model there is no God and no consequences for mans actions. No definitive infinity would have no laws of structure. No laws of structure and you have no definitive values of physics. Thus Stephen Hawking writes "the uncertainty principle places limitations on the accuracy of all of our predictions". Actually the uncertainty principle, as with infinite numbers, proves the existence of an infinite creator that forced it all into being. The uncertainty is the infinite value that an infinite creator left as a clue to his existence. For that cause, one can never know the exact size of the infinite universe because once it's measurement is calculated at any finite point of expansion, you must always add the factor of it's expansion in the time it took you to calculate it's size. In this way to calculate the size of the universe is an infinite task constantly needing correction. What the E- plain does is just as Stephen Hawking writes, "what the singularity theorems really indicate is that the gravitational field becomes so strong that quantum gravitational effects become important: classical theory is no longer a good description of the universe....There need not be any singularities in the quantum theory".

But also the opposite is true in that our universe has it's

own singularity. What I mean to say is that the finite matter and energy in the expanding universe is balanced with all the matter it can contain and maintain expansion. This universe may not lend matter or energy with out gaining matter and energy from some other universe to maintain that balance.

If figure 1:2 is the case then the energy exchange happened in the past and is connected to the present sharing of parallel universes matter and energy. E would still = MC squared but as follows: $E1=M1 \ C1^2$, $E2=M2 \ C2^2$, and the same for universe 3 as $E3=M3 \ C3^2$. The formula for the E-plain would have to be able to maintain the C^2 of the constant in all three universes. E-plain Energy and Matter would have to be constantly increasing only if matter in the universe was increasing matter and energy. With E and M constant then the gravitational pull of expansion would never stop. As matter moved farther apart the gravity of the E-plain would move outward ahead of it like magnetic fields. Since the matter of the black Holes in the universe are so strong and dense, their fall would never be greater than all the energy and matter in the universe. I suspect that there is maintained just enough to keep things expanding at a constant rate with out much slow down once the balance of inertia is in place being feed by the black holes that are newly created from supernovas. The E-plain would also explain how new stars are able to be born. The gravity of the E-plain would force matter and gases together at some point of expansion and transfer the energy required to make a new star. Gravity now becomes a constant in the universe also. But in two parts, internal gravity and external gravity. You certainly can not take something away from one side of the equation and not add it to the other. It is elementary algebra and Uncle Albert's constant.

In figure 1:1 we see time in 3 dimensions being Forward, Width, and Depth. Time past is not drawn into this figure but would be a reverse arrow moving toward creation. Recent theory predicts that super-strings can be utilized to travel back in time. But to do so you must escape the effect of the 3 dimensional aspects of time in the event horizon of now. The E-plain has its own time and is the vortex to the past that allows super-strings to connect us to the past. If we could use a mono-pole superconductor as an attract-er to tune into the energy frequency of the super-string then perhaps we could use the magnetic field of the E-plains gravitational force to pull us into the past at the speed of light. Perhaps these strings of energy can be mapped at given equal values and traveled upon through the universe that would make light speed achievable by using the map to avoid collision with matter in the universe. Matter would perhaps reflect the signal of this E-plain like sonar reflects sound waves and sensors could detect it far in advance and prevent a spacecraft from disaster. Such a device could use several of these star-drive engines to slow speed and navigate between planets and stars.

By riding energy maps and using the power of attraction rather than propulsion we could enter the 21st century as the first generation to prove that warp and possibly time travel is more than science fiction. It is now science fact that the magnetism of gravity and energy in the E-plain allows. Black holes would most likely reflect this electromagnetic field into the universe (at 3 degrees K). If we used the frequency of the pulsar to hear it's signal with radio telescopes then perhaps we should look in the electromagnetic spectrum for other types of dead stars. Our Earth is a giant magnet that the magnetic effects of our sun pulls upon to keep us in it's orbit in the same way subatomic protons, electrons, and other particles are holding matter together with negative and positive charges.

Magnetism gave use electric and light and taught us to move electrons through wires. Lasers have shown us how to use light to hear sound and how to even destroy bullets as they fly through the sky. Now we should see if gravity is pulling light into a Black Hole because within it magnetism and gravity are unified. The result must be the speed of light in my view. Any matter, wave, or particle traveling at the speed of light must appear as light to the stationary body. Since the gravity of a black hole warps light into it, then it only stands to reason that to harness it's attraction qualities would create a craft that could travel at the speed of light. The E-plain must some how to unify the forces of gravity and magnetism somehow to do this. Perhaps the E-plain is only using gravity to attract matter and magnetism to move matter farther apart leaving an ever expanding universe constantly moving energy from one edge to the next. This is within the E-space within the fabric of all matter and energy in space.

The question of life arises in this model. The principles that are generally accepted must meet Feynman's proposal, Euclidean space-time calculation, and Einstein's curved space-time where particles try to follow the straight path in curved space and appear bent over histories of the universe. Finally this principle must explain life it self and how it formed on a subatomic level. With this cone shaped growth of the universe we would

25

meet the test. The universe is in a clockwise rotation at the point of creation with this model. The growth becomes a spiral as reflected in DNA strings and it's clockwise spiral. No left-handed DNA would appear in this universe. The particle would appear to be bent due to the expansion causing it's path to form a spiral. The universe could then be a sphere shape of finite size much like a scope of ice cream at the end of a sugar-cone. Our ice-cream here would be growing with expansion and the rotation of the ice-cream would throw off the background radiation that we see at a certain phase and 3 degrees Calvin. The spiral of the E- plain would then be pushing against the universe and forcing it up and out in all directions. Space-time is being created in all directions. From the big push or bang, this spiral route expanded like a bubble on a twister ride.

Time in our universe is then forced into one direction while the time in the E-plain is forced in the opposite direction. Much like a whirlpool the energy is going in a circle that forces buoyant matter away form the center of the vortex. Since energy is converted back into the universe there is no need for the temperature to cool with expansion. Particles in space would reflect that background radiation at it's reflected temperature at which expansion is maintained. As with Einstein's theory we have a constant of energy and matter in balance of the universe in expansion and the E-plain of Black Holes energy moving with it into an infinite motion. The spiral will never crunch in this model. To do that it would have to meet a brick wall of anti- matter an become non-existent. Anti-matter can only exist in the E-plain at the point of creation and would then decrease to a constant of matter heavy enough to continue the spiral motion of the universe or universes it forces into expansion. God can then place it all in motion and then stand back and watch it grow. Genes and there model as we now can view it under an electron microscope would reflect that spiral, and they do. Particles would also spin in balance with the force of the E-plain or the clockwise direction of the known universe. Anti-matter proves this. If we can reverse the spin and see anti-matter in a particle accelerator, then it must be in the frame of quantum mechanics somewhere. This is the E-plain's tool in the first moment of creation to create the gravity and motion needed to begin the spiral. After the anti-matter has completed it's attraction force needed to form hydrogen and the mass of matter then it, would decrease and the stars would take over to form the energy needed to continue expansion. Is this not what we see in the cosmology we know? Therefore the God-particle is just the outward remnant of the frequencies of the voice of God himself using sound to create.

So we can see a universe or universes pushed into infinite expansion and propelled into space and time by the direction of attraction first by anti-matter, then by supernova stars. If the big bang was so hot as we read then anti-matter would be replaced by the effect of stars in a very brief moment. Black Holes would soon after come into being when the matter had sufficient space to maintain gravity and the spin of the universe. Inertia would force out as the E-plain would push in at the base of the vortex and expansion would be procreated by the stars and the Black Holes that come from them. For every force there is and equal and opposite force. That force can be seen in the stars and the infinite expansion of space and time. Matter and energy in this universe must be maintained constant for this to work. Space and time must be unlimited to keep it in motion. Thus we have an E-plain of antimatter in the first moments of creation and then a balance of it's properties in the eventual force of it's existence. The voice of God spoke and that became energy within a unity inside matter in this E-space.

The E-plain would then have an E-gravity, E-time or God-time (Quantum time), E-space, and E-matter which is anti-matter that is replaced with super-dense matter. It will also have E-energy directed at the universe where light could not escape. It's field of energy must pull and push everything including light in order to work. To maintain the constant it must also reflect it's energy into new stars and matter to continue expansion and adjust for increased size. And since there is no new quantity of matter and energy, it's work would never have to push or move more matter than it balanced to push after stars came into being. It would and does only effect matter in the event horizon of the Black Hole stretching it into a gravity of quantum magnitude to attract the matter and cause space to expand. It is much like being in a balloon at it's exterior surface, filling it with air, and moving yourself further away form the center with the more air you fill it with. Space and time is the air that the E-plain expands and matter would be the substance of the surface of the balloon which remains the same quantity but moves with the expansion.

As for the mathematics of this equation it is beyond me. I have tried to present my theory in a way that will help some other physics math whiz to achieve answers to those calculations. I think of this universe and the effect of a Black Hole as a piece of matter on the surface of space, so heavy that it pulls space and time in all directions and drawing matter and energy along with it from a distance. This super- heavy piece of matter which was once a star will not stop it's fall until it falls back to the point of creation or hit's an anti-matter universe in a parallel dimension and disappears. It may also collide with another Black hole in another universes' E-plain that is shared and explode reflecting it's energy back at the universes. The only other possibility is that it never stops its fall and the universe expands forever by the attraction and forces of the E-plain or the Black Hole itself. To me it seems that things that occur here on Earth reflect what happens in cosmology, as with hurricanes and anti-cyclones, South and North poles, and magnets. So it is conceivable to me that an anti-matter universe may exist. If this is so it would demand that there be some force and quantum machine to keep it separate from the matter universe. The two would have to have been created at the same time and their spin would effect each other to force them apart. The forces between these universes must have an energy exchange point much like that of two children on a school yard pushing each other and falling away from the center of their force.

With this model a singularity and a quantum can coexist together and so can God who pushed it all into motion in the beginning. Life can also exist in another universe but it would be as unique as we are and follow the same laws of physics that Einstein predicted. For that life to reach another universe it would have to travel at the speed of light in the E-plain and have control over navigation to place it at the exact time and place of space-time desired to arrive at the matter (planet) of it's destination. If warp 2 or 3 is possible (2 or 3 times the speed of light) then this could only occur in the E-plain or by harnessing the force of attraction of the E-plain. This idea suggests that in order to travel through dimensions one must have control over time and be able to warp time. Such a device must use the laws of physics to do that task at hand. Relativity tends to lean in this direction of thinking with this model and the picture we view is one that might explain why biology maintains age at a specific gravity. That is to say, that if we lived on Pluto would we age 247 years or one year. Clock experiments have proven that time is relative, but is biology? If it is relative in the same relation as theory seems to allow, then the universe is all the more grand and may be traveled by escaping the bounds of the direction of time we travel. Thereby we can with more ease overcome the vastness of space. After all it really does not matter at which time in history past we arrive at another planet, only that we arrive some time after the big bang. It is safe to assume though that any life form arriving at another planets timeline would become a part of their history. So the resulting question is here can sound travel faster than the speed of light. Can Frequencies at that speed create energy and matter within this E-space and create strings that vibrate and separate the parts of energy which makes up all matter.

Such an act of just going there through time would be an act of cosmic violence against the biology of that dimension. To risk the natural progression of history and science would be a crime. Even more so to risk the progress of biology would also be a momentous crime that could have destructive results. Just because we can travel through time does not mean we should. Perhaps that is why we have trouble seeing UFO's and documenting there existence on a mass media scale. We are perhaps visiting ourselves from the future and doing all we can to keep from upsetting the time line with a paradox. To do that we must hide ourselves well form our past to travel there. The prime directive of StarTrec becomes a very important idea that it's creators gave us as a legacy to science, that we not change the progression of events past and alter the future. Not on Earth or on another planet in our universe or some other universe. To do such might be effect enough to place the fabric of existence into jeopardy. Think of what a mess that would be for God to clean up.

We have looked at the brief moment of creation by seeing it's effect on the present from our point of view in space time. Einstein was proven right again and we attempted to unify the forces that cause the universe to come into being. In a way we have looked at creation as an ongoing process that is still expanding into eternity. We have also given a possible explanation on why God exists. Even with a multi-universe model we still need to know what the force was at the moment of creation. All information seems to point to an infinite force with an equal amount of power and strength to push an infinite expansion. The E-plain within this vast infinity must also be a spiral of time as our expansion is. The sphere shape of the universe is expanding in all directions and rotating around the center of force in the E-plain. Behind the quantum machine is it's maker, the hand of God. This fractal shape of space we live in is his signature on the very parts of creation that reflect his name.

Einstein found that matter and energy were related. He also found that time and space were related. I have tried to show, with my limited understanding of math and formulas, that the time of matter and spcace-time are related. That the energy of magnetism and gravity are related. That the age of the universe is not the sum total of all it's pieces of matter. That matter-time is warped by

31

energy in the universe. I tried to show how science in this field has made errors in trying to date matter by energy and size. And doing that is like adding up my age by adding the total age of the energy of each atom in my body. That would most likely put me at the age of a trillion years old. The general theory I have presented here is that the growth of the universe is like a cone with Black Holes pulling it into expansion by an energy field that is reflected in the universe and it's matter. Space tells matter how to move and matter tells space how to warp. Energy tells matter how to reflect time and matter tells energy how to expand time. Matter rotates in one direction and anti-matter rotates in the opposite direction. To travel back in time the field energy of anti-matter must be used to attract us in the opposite direction of time and matter. To travel at the speed of light we must ride the unseen force in the universe I call the F-plain, and use it's attraction to pull us at 186,000,000 miles per second. And as matter warps space into expansion, it reflects the eternal power behind the universe. That power is Yah Vah God.

" By Messiah All Things Consist" The Holy Bible

UNIFIED THEORY

$$E = M\ 12f^{12}\ C^2$$

$$12f^{12}\ (Hz \sim MHz)^2 \quad v = f\lambda$$

$$Et = Mt\ C^2\ t$$

$$E = 12f^{12}\ \lambda\ C^2\ t$$
where v is the speed of the wave (c in a vacuum, or less in other media), f is the frequency and λ is the wavelength.

And this time. As waves cross boundaries between different media, their speeds change but their frequencies remain constant.[1] 12 dimensions with string packets in themselves vibrating in harmonic and anharmonic frequencies creating energy and time-matter.

Sound is in Hz or MHz, hertz or megahertz, and it has been discovered that when radio frequencies are bounced against each other using tuning devices, that radiation is created at given levels that can be used in modern medical equipment to cure cancer. Think of it as acoustic energy that is created by wavelength or particles within strings that vibrate at a given range of energy the same way you tune a note to by "C" on a piano. When you play that note along with other possible notes, you can increase the strength of the harmonies to make a cord. If you could then make the cords play with opposing energy waves then you create radiation which vibrates strings. The strings are long groups of particles that combine to make up a displacement of space-gravity that is vibrating within a range of radiation or energy. Some of the strings will be more dense than others, yet each will put off a spectrum of energy you hear, see, touch, feel, or can be united to form matter, energy, light, and space.

Albert Einstein proposed that light quanta be regarded as real particles. Later the particle of light was given the name photon, to correspond with other particles being described around this time, such as the electron and proton. A photon has an energy, E, proportional to its frequency, f, by

1   https://en.wikipedia.org/wiki/Electromagnetic_radiation

$$E = hf = \frac{hc}{\lambda}$$

where h is Planck's constant, λ is the wavelength and c is the speed of light.[2]

Think of it as a jump rope that is attached at both ends to a voice that causes it to spin round so fast that the arc of the spinning rope appears solid. When you touch it the space inside the spin is not known or seen, but the surface is solid and when you touch it has qualities unique to the type of vibration it holds in frequencies.

On our piano we have 144 notes and 144 thousand combinations of notes that each make up different songs. These are the energy we see when we break atoms. When we look inside the spinning rope of the strings we find the energy like a camel center to a chocolate shell of matter vibration in time.

Now when we bombard the sounds within the space-gravity inside the spinning string, our jump rope, we find that it gives off energy in magnetic fields, radiation, and gravitation that warps the space inside into energy and the outside into space-gravity as Einstein's relativity has shown. The warp of the light around the sun which proved relativity also proves that the relative vibration inside string-space, not seen by matter on the outside, is warped to a frequency of energy which is unified in every part of matter and space.

We have 12 to the 12th power of possibilities n frequencies within the spectrum of energy that can make sound waves

---

2   ibid

and radio frequencies combine to make energy. The answer all along was in the name and in the bible. "The Big Bang" is Sound that has a frequency that was discovered by the Bell labs at 3 degrees Calvin.

"sound traveling is not isothermal, as believed by Newton, but adiabatic. He added another factor to the equation- gamma-and multiplied $\sqrt{\gamma}$ to $\sqrt{\dfrac{p}{\rho}}$, thus coming up with the equation $c = \sqrt{\gamma \cdot \dfrac{p}{\rho}}$. Since $K = \gamma \cdot p$ the final equation came up to be $c = \sqrt{\dfrac{K}{\rho}}$ which is also known as the Newton-Laplace equation. In this equation, K = elastic bulk modulus, c = velocity of sound, and ρ = density. Thus, the speed of sound is proportional to the square root of the ratio of the bulk modulus of the medium to its density.

Those physical properties and the speed of sound change with ambient conditions. For example, the speed of sound in gases depends on temperature. In 20 °C (68 °F) air at sea level, the speed of sound is approximately 343 m/s (1,230 km/h; 767 mph) using the formula "v = (331 + 0.6 T) m/s". In fresh water, also at 20 °C, the speed of sound is approximately 1,482 m/s (5,335 km/h; 3,315 mph). In steel, the speed of sound is about 5,960 m/s (21,460 km/h; 13,330 mph). The speed of sound is also slightly sensitive (a second-order anharmonic effect) to the sound amplitude, which means that there are non-linear propagation effects, such as the production of harmonics and mixed tones not present in the original sound. " [3]

Within each string packet inside the spin is a unity of energy at the level of opposing harmonies that have this anharmonic effect of energy.

The modern theory that explains the nature of light includes the notion of wave–particle duality. More generally, the theory states that everything has both a particle nature and a wave nature, and various experiments can be done to bring out one or the other. The particle nature is more easily discerned using an object with a large mass. A bold proposition by Louis de Broglie in 1924 led the scientific community to realize that electrons also exhibited wave–particle duality. [4] This is in fact because under the surface of the vibration of strings is the center of frequencies from sound that cause sub-particle vibrations. When it tune one matter becomes Iron while another become Uranium and so on.

"As a result, oscillations with frequencies 2ω and 3ω etc., where ω is the fundamental frequency of the oscillator, appear. Furthermore, the frequency ω deviates from the frequency ω_0 of the harmonic oscillations. As a first approximation, the frequency shift $\Delta\omega = \omega - \omega_0$ is proportional to the square of the oscillation amplitude A:

$$\Delta\omega \propto A^2$$

In a system of oscillators with natural frequencies ω_α, ω_β, ... anharmonicity results in additional oscillations with frequencies $\omega_\alpha \pm \omega_\beta$.

Anharmonicity also modifies the profile of the resonance

---

3   https://en.wikipedia.org/wiki/Wavelength
4   ibid

curve, leading to interesting phenomena such as the foldover effect and superharmonic resonance."[5]

This all happens inside the strings which I suggest are in fact not just 11 as modern string theory suggests, but in fact 12 with the inner dimension of vibrations within the displaced space-gravity inside the spinning or vibrations beneath the surface of matter. Matter-time and Energy-time are thus reflected and absorbed as needed to give off heat, magnetic effect, gravity, and radiation as observed in the modern device used to treat breast and prostrate cancer developed by an amateur radio inventor in Pittsburgh to treat his own prostrate cancer. [6]

What Kanzius discovered was in fact what is inside every string packet in sub-particle vibrations that causes matter to displace space, and causes energy to be created within those packets below the surface. We observe the surface as Matter or watch it convert into energy as we burn it for fuel or split the packet to cause nuclear explosions.

My crude attempt at physics mathematics may show in this but I am sure that the vibrations of frequencies can create energy and that vibrations can displace space with matter. The spectrum of energy may vary to include more than just sound, but sound is at it's core because of this simple truth found in the first words of the bible:

***Genesis 1:3 And God said, Let there be light
When God SAID he sang the vibrations that created
the sound of the big bang that displaced space with
matter and time, energy and the universe moving
faster than the speed of light.***

5   https://en.wikipedia.org/wiki/Anharmonicity
6   https://en.wikipedia.org/wiki/John_Kanzius

So matter absorbs energy that reflects a greater time than the actual time of matter because it has been traveling across space at the speed of light. To carbon dating it appears millions of years old, but like the 2 month old skin cell in your body it reflects the age of the whole body in space-time within the universe that is you. The cell is less in time than the whole. Empty space is warped by matter and matter is warped from inner-space by unified vibrations which reflect energy. When sound vibrates strings space moves. The electromagnetic and energy effect is greater inside the sub-string inner-space where foldover effect and superharmonic resonance creates like on a piano of 144 possibilities of notes of sound. The harmonies and energy effects of $E = 12f^{12} \lambda C^2 t$ with energy creating matter from inner space where 12 strings play the voice of God. The counter effects of the $f^{12} \lambda$ working against each other form the unity of all forces to tune in the right f that displays matter in different types to our universe. With 12 strings, we can play 144,000 combinations of what is creation. Each part of the strings can play cords or single notes that make unity of force divide like a spectrum out of the prism showing energy, light, and matter in its visible parts. The Unity is within, not without as light splits into color with the prism. Frequencies that create by opposing forces inside my E-space packet.

RATE TIMES DISTANCE

So when factored into the relationship of rate times time equals distance we have the formula that proves the expanding universe is only about 6000 years since creation even thought the matter may reflect the older age of expansion. We can further see in present physics studies that the unseen energy that is called "Dark Matter" is believed to be a source of force of gravity that has yet to be discovered and remains in theory alone the idea of a universe without God that by this substance could be an explanation of what is causing expansion. So we must move on to other reasons to believe that will further explain why the earth is only 6000 years old and was created in seven 24 hour days as the Bible says in Genesis.

A theory of the Evolutionists is that the earth is millions of years old and that it has somehow come from nothing created by nothing when it exploded from a cause unknown. Yet we can not observe this in any form in science or in the scientific method due to the law of causality. Something can not come from nothing. For each action there is an equal and opposite reaction.

So let's entertain the idea of the limited resources of earth and go back just a few thousand years that the evolutionists suggest in Genealogical science and take their model of known mankind from the 50,000 year mark given at Family Tree DNA. Let us see if there theory works.

Now these evolutionist Genealogists agree with the Bible in that they have traced the DNA code and found that at one time there was according to their model and theory just one man and just one woman. They even call these two by the names that the Bible calls them, Adam and Eve.

Now they also suggest that they have traced back the MT DNA and found that at one time, they say about 20,000 years ago there were just four women. They say that every person holds this DNA in there code and can be traced back to one of just four Great Grandmothers. Yet we in Scripture studies knew this from the story of Noah and the

Ark. Ham, Shem, Japath, and Noah's Wives were all that was left of ladies after the flood and therefore the Mt DNA would reflect that.

So let's do the math and see if this would work.

Now we know the generations of mankind are by the life expectancy of when a child can be produced biologically from a woman and when a man could mate with a woman to cause her to have a child. After all this person could not have just appeared from nothing. So we look back at the age of woman without modern science and medicine and see that at the time of Christ a woman lived to be about 40 to 50 years old. Now some may and did live longer, but many died in childbirth and therefore the harshness of life had it's toll on people so a generation would be about 50 years.

The child that grows to about age 13 would then be biologically able to have children and give childbirth and the next generation would by about age 13 able to then continue the process. The combine time is about 26 years and then the next generational birth cycle about 39 years giving the grandmother the ability to see her great grandchildren. For some no doubt the process was a bit longer so that by age 20 a new generational cycle would begin. Then to age 40 to the next set of parents. Interestingly the Bible puts a Jubilee or a generational cycle at 50 years which should be enough to see at least 2 generational cycles from Father and Mother to Children.

So let's use the 50 years cycle and take our population growth factors into accounting and come up with a number that should be the total population of the earth given other factors such as war and sickness.

We presently have 7 billion people on the earth. We know that the Plague and War from the dark ages to now have killed millions of people yet some were not effected and still had children that went to the next generation. Yet we see no major event less than 6000 years ago that has proven that any major event happened that caused a mass culling event other than what is listed above, at one time, there were just four mothers left.

So let's take that evolutionist date of 20,000 years and divide a generation and calculate a factor of just 4 children each cycle because there was not such thing as birth control back then. 20,0000 Divided by 50 = 400 generations. Now let's calculate the factor of exponents of expanded generational increasing population.

X^2 multiplied by 400 generations = 10,240,000,000,000

people or over 10 trillion people.
The land mass of the earth is 57,505,693 square miles in area. Dividing by the population above we get 178,069 and one third person per square mile. If this were so then you could not walk one mile without running into 178 thousand more people. In farming you need in modern production 7 acres of land to make enough food for one person. Each acre needs 20,000 gallons of water and each person needs about one gallon a day to drink to stay alive. United States Geological Survey Oct 5, 2015 - Notice how of the world's total water supply of about 332.5 million mi3 of water, over 96 percent is

saline. Of total freshwater, over 68 percent is locked up in ice and glaciers. Another 30 percent of freshwater is in the ground. The total numbers do not add up. If even there had been on fourth of what the evolutionist claims then the earth would have run out of food and water long ago at just 20,000 years of population growth. So when we factor in for hundreds of millions of years then we still with mass extinction events would have to replenish water and resources millions of times over to accommodate even the simplest of life forms with animals and plants and insets. Yet all these billions of creatures remain here today to prove that evolution is more of a religion than creation is, for it would take much more faith to believe that the whole magic water replenishment would have kept going in light of these numbers.

Now when we take the 120 generations which is 6000 divided by 50, we get without any outside factors about 20 billion people given just an average of 4 children per couple in population growth. With the previous generations passing every 50 years we get a total earth population of about 7 billion which is what we now have in the year 2015. So the real number of people that have lived since the flood of Noah is 20 billion and that number is just over about 4000 years ago, not 20,000 that evolutionists claim. In the previous 1000 years since creation the population would have been about one fourth of that or 10 billion that died in the flood. We then have the total of mankind at about 47 billion since creation.

So now we have proven by scientific method and observation of the actual results of the theory of evolution and it's effect on population and the resources of the earth and proven it wrong. The earth could by the formula of rate times the time equals distance alone prove also that the expanding universe is only 6000 years old also. Then we have the solar dust on the moon that proves that it's rate of accumulation is only about 6000 to 10,000 years old. The rate of H3 or Dust from the Sun collection on the surface of the moon was calculated and if it had been over millions of years old then it would be thousands of feet deep. In 1969 when man landed on the moon the dust was only a few inches deep. In fact, there was suggestion that if the surface was soft dust that the lunar lander would have to reverse engines to avoid being buried under the deep dust which could have stranded them on the surface of the moon. These FACTS known by those of us who watched the TV broadcast live as children alive still today knew this and heard the newsmen talk about this possibility on TV.

Finally we have the fact that there is no fossil records or evidence of any transitional forms of life that ever changed from one species to another. Nor have we been able to create any natural and spontaneous forms of life that transition between plant to animal or from an animal in one species to another.

Nowhere has any living changing evolving being been found, and biologically science tells us that mankind can not procreate with other animal spices let along cross migrate DNA into other forms of life except through procreation by sexual methods with like kinds. There are no monkey-men or bird-frogs or tree-lizards and none have even been found because the code in our DNA prevents us from procreation with other species of life. The Bible says this, everything was created unto it' own KIND.

We have also proven that the voice of God caused creation. Deep inside the string of His music is unification it self that Ernestine searched for.

So when we look at the observation of what we can see and what we know by evidence of what really is science, we find that the evolutionary religion is just that, a FAITH in a belief that there is no God even in the face of overwhelming evidence that God created everything.

So when we of Bible FAITH that believe in God see such idiots as Conan O'Brien mock those who believe the earth is just 6000 years old on TV we know that he is the product of brainwashing from public schools that demand that we agree with the religion of evolution that denies the scientific facts of the created universe. You see they were foretold in the Bible in prophesy also, that this time would be filled with mockers walking after their own lust of the flesh that would deny the very creator that made them.

The Bible also says they would be full of sin and lawlessness and that they would worship the creation above the creator, and that they would be ever learning of things with knowledge being increased, yet they would be blind to the truth of Yahuah God and his having created everything.

Now also there are those Christians who claim to believe in God also, yet compromise with the theory of a large timeline of millions of years who claim that the 6 days of creation happened over thousands or millions of years also. Yet we know that man who breaths Oxygen and Plants that breath Carbon Dioxide had to be created during a very short period of time in order to maintain symbiosis of life and species. So the compromise is also wrong and an attempt to blend the profane false with the sacred truth.

When you agree with the world that denies God you compromise with a lie that imagines that the world is somehow able to come into being when nothing exploded caused by nothing and a DNA code dividing species came from a one cell organism when it was created from rain on the rocks and somehow the Frankenstein electric shock made life come from death and matter that can not move or walk like an idol of stone or wood. Life can only come from God and the code of Life from his design and creation that must have an author of the DNA code and the forces of energy in order to maintain laws by which we are able to exist. Just think if there was no law of Gravity, or law of Energy such as $E = M C^2$ We must conclude that the only way life and the universe can exist is by a creator who made these laws to govern how particles and planets move in the expanding universe. What scares the evolutionist is that if this is true, which it is, then the other Laws found in Exodus 20 as a moral code is also a law they can not escape. That means some day we will all face the Creator and answer for what and who we believed in.

EVOLUTION IS RELIGION

– 4500-year-Old Proof at Hell Creek Dig shows Triceratops Foot print and MANS FOOTPRINT side by side, at same level, and inside. Evolutionists want you to believe that man came from apes. Yet their religion can't explain this one! Triceratops was on the evolutionary calendar in the Cretaceous period 125 Million years before man. So how could man, who didn't evolve yet, run with the Triceratops, fall and leave a hand-print, and step into a triceratops dinosaur footprint?

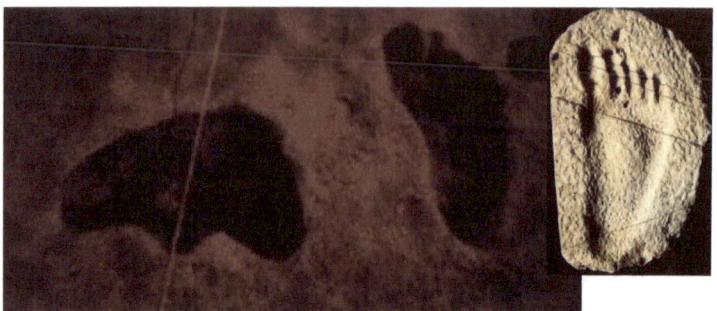

**SO WHICH RELIGION DO YOU BELIEVE?
EVOLUTION?**

**Man's Hand dates back at the same level to the
Dinos found at Hell Creek Dig.**

Dino and man walked together about 4000 years ago.

Hell Creek Dig Takes Imprints of Dino and Man

It never seems to lose the feeling of shock how some so called educated people can allow themselves to be so easily tricked into lack of reason with such an abandon of all powers of reasoning out the truth. The world is filled with the folly where blind men read to blind men about things that just never happened.

A good example of this is found in the religion of Evolution. Science as it is taught to our children today in the institutions run by the high priests of the evolution religion claim that their faith which they have not seen or proven is true because the absence of evidence proves it is true. This circular reasoning is mind-boggling with stupidity once you truly reason out what they are saying. They claim that evolution took millions of years for a monkey to become a man. Therefore if their claim is true, since we have been around for millions of years evolving

48

from a monkey to a man, then the theory that evolution happens should have made all the monkeys into men by now, or at least have some life forms in transition somewhere in the middle of that change from monkey to intelligent human beings. You would think that reason would prove that the climb up the evolutionary chain would have to continue to happen if it ever did happen, because without a God or Supreme Being to write a code into DNA that would prevent any species from changing it's form, all would naturally occur as it had from the beginning of transition. Next you would expect that any logical thinking person would be able to see that the most desirable state of any animal would be to evolve to the highest level of the evolution process and that the fact that we are here as humans would have over the millions of years caused all animals to achieve the highest form of life possible. Therefore if evolution were true, then we would all be the best humans that 125 million years could produce and there would be no monkeys left to evolve. So the religion of Evolution in their lack of evidence and demand that we worship nothing or be burned at the stake for heresy against it's sacred priests called professors when presented with their circular reasoning which is intellectual folly demand faith without evidence and have even over the years fabricated evidence that was not there by just telling lies.

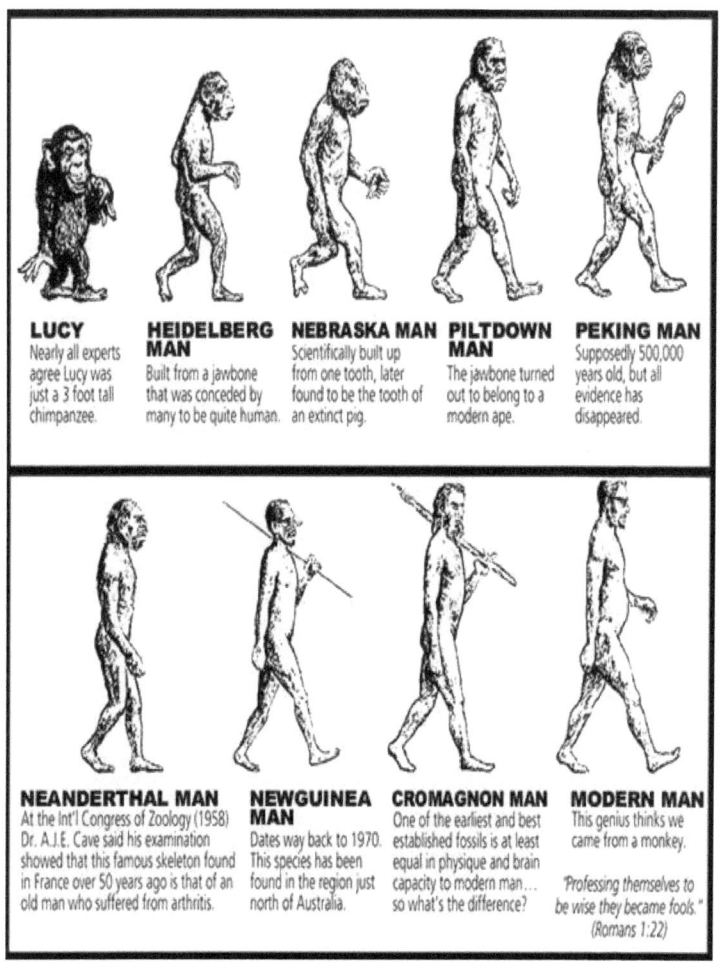

LUCY
Nearly all experts agree Lucy was just a 3 foot tall chimpanzee.

HEIDELBERG MAN
Built from a jawbone that was conceded by many to be quite human.

NEBRASKA MAN
Scientifically built up from one tooth, later found to be the tooth of an extinct pig.

PILTDOWN MAN
The jawbone turned out to belong to a modern ape.

PEKING MAN
Supposedly 500,000 years old, but all evidence has disappeared.

NEANDERTHAL MAN
At the Int'l Congress of Zoology (1958) Dr. A.J.E. Cave said his examination showed that this famous skeleton found in France over 50 years ago is that of an old man who suffered from arthritis.

NEWGUINEA MAN
Dates way back to 1970. This species has been found in the region just north of Australia.

CROMAGNON MAN
One of the earliest and best established fossils is at least equal in physique and brain capacity to modern man... so what's the difference?

MODERN MAN
This genius thinks we came from a monkey.

"Professing themselves to be wise they became fools."
(Romans 1:22)

Source Jack Chick Publications

It takes more faith to believe we came from apes in absence of any fossil record or transitional forms of life than it does to believe in a creator who made use with a genetic code that prevents use changing our created form. Further more this trick of no evidence being evidence that the impossible is possible and creating false evidence to prove lies as the truth is a trick invented in modern education by the Roman Catholic Church. The church in fact did all it could to prevent people from reading the truth in the Bible for thousands of years because it would expose their lies about how Yahshua died on a stake and not a cross, and show how they in fact they erased commandments on idols so they could sell idols and icons to the people. Further more it would expose that the Church had been telling lies for centuries about being forgiven by making payments to the church when you sinned at the church run whore houses. So before you jump to the quick this Unleavened season and forget to read the truth found in the Bible because of the circular reasoning of the tricks played on mankind by the Romanized Lies of Jesuitry, you might want to take a look at hsitory on how the Bible got to us to read, and how many died to bring this to you.

For like Evolution has sold it's religion without proof, the Roman cult of Popes and Kings did all they could from the dark ages to keep you from knowing the truth and exposing their lies as what they are, circular reasoning with no evidence of any religion of idolatry found in the pages of the Truth, the Bible, as coming from God.

How do I know Evolution is a lie? In one word

SYMBIOSIS

symbiosis meaning an interaction between two different organisms living in close physical association, typically to the advantage of both.

A mutually beneficial relationship between different people, species or groups. Living organisms dependent on other organisms in order to maintain life.

Trees and plants breath in CO_2 and exhale O_2 or oxygen cleansing the air we breath as animals. As we breath in O_2 and exhale CO_2 or carbon dioxide we are giving life's breath to plants and trees.

IF one was created before the other or evolution was true, then no human or plant life would exist on earth. Both in symbiosis had to be created and come to maturity simultaneously or the others could not maintain life. If plants had come first alone, then the O_2 would have choked the life from all plants and they would die under their own oxygen rich environment. All the volcanos could not have put out enough gases to keep the trillions of plants alive, and in a heavy O_2 environment the earth would have burned up at the slightest flame from lightening or a volcano's flames. All the plants would have died in a world on fire.

Likewise if man and animals were created first or evolved over billions of years as evolutionists claim, then all animals would have been suffocated in CO_2 and the resulting environment would have been a barren world void of life.

Yet in this world we live in, there are so many species of insects, animals, and now over 7 billion people who all have symbiosis with each other in many different ways. Worms and insects break down soil and provide nutrients for plants to grow and give us fruits and vegetables that we eat. Animals and birds give nutrition also to the soils and help clean off pollution through the break down of chemicals into soil. Algae feed fish deep in the oceans, and the environment supports all life in symbiosis.

There is no other way that life could exist without a God who created it all in a very short span of about 6 days as the bible says. The very fact that we are alive means Yahuah God is also alive, and He created us all.

The religion of evolution is dead. It's preachers are left in empty halls of self impressed theories that have no fossil records to prove anything they say. Every DNA test done on everyone of their so called fossils has proven they have made up the whole story of evolution to appease themselves of the guilt of their own sins. They do not want to face the facts that Yah is God and He is not dead. If they do, then they would have to accept the moral imperative that all mankind will be judged in the end.

You see, If there is no God, or if He was dead, then there would be no moral absolute in what is righteous and what is unrighteous. Without any law or moral code, then mankind would be free to commit chaos. Anything that any one would do or say would not mean anything, nor would any law of morality have any purpose. This is why evolution at it's very core is a chaos theory. It claims that rain on rocks mixed with a chemical soup to form life from nothing without any limitations of what kind of life it would become. Yet it fails to explain where all things came from or how they got here to allow for this random encoded event of different species.

So if evolution were true, then the unlocked code of creation would be free and open for any strange form of life to come from any other. A tree could mate with a human, a bean plant with a bird, or a fish with a lizard.

If all life came from one organism, then all organisms would be related and could and would have means by which we would all have procreation or some sort of means to propagate more life and species. Yet the fossil record and the observed science of today prove that this is not the case.

Some intelligence has locked the code of life between species so that they all can only procreate their own kind. Not cell ever produces any other kind of plant or animal. In fact, for over 150 years of science and countless tests in labs by thousands of scientists not one form of life has even been created from a chemical process alone. Nor has any scientific method proven that anyone can create life except through sexual procreation and only by the means to make life from a life.

So in truth, evolution and the religion of atheists is dead. There must be a creator or else there is no life. Trillions of stars and galaxy's as well as billions species on this one planet earth prove that their is a creator.

Now about 2000 years ago Yah God sent His own son into the world, that who ever would believe in Him would not parish in the end but would have ever lasting life. He died for your sins, shed His blood to cover your sins and by the commandment of Yah God gave His life in punishment for our shortcomings. Then three days later He arose from the dead. Then He promised He would come back again at the appointed time to bring all those who love Him and who believe in Him to live with Him forever in perfect sinless bodies that will never die.

The results of sin is death, but the gift of Yahuah God is eternal life through Yahshua Messiah our Master. This is what the word of YAH GOD who IS NOT DEAD says. It is His love letter and commandments to us while we live, to honor Him and worship Him alone. There is no other way to have eternal life. There is no other way to escape punishment for sins. There is no other way to life but through the giver and creator of life and the universe, YAHUAH GOD and HIS ONLY BEGOTTEN SON, YAHSHUA MESSIAH.

John 3:14 - 21 *"And as Moses lifted up the serpent in the wilderness, even so must the Son of man be lifted up: That whosoever believes in him should not perish, but have eternal life. For God so loved the world, that he gave his only begotten Son, that whosoever believes in him should not perish, but have everlasting life. For God sent not his Son into the world to condemn the world; but that the world through him might be saved. He that believes on him is not condemned: but he that believes not is condemned already, because he hath not believed in the name of the only begotten Son of God. And this is the condemnation, that light is come into the world, and men loved darkness rather than light, because their deeds were evil. For every one that does evil hates the light, neither comes into the light, lest his deeds should be exposed and corrected. But he that does truth comes into the light, that his deeds may be made manifest, that they are works begun in God. " Yahshua Messiah.*

**This Fact leaves you with one final question:
Do You Believe?**

If so, then you might try talking with God in prayer and inviting Yahshua Messiah (Jesus Christ by his Hebrew name) to come into your heart and forgive you of your sins. To write your name in the book of life of those who believe on him, and the book of remembrance of those who call upon His Name. The bible says that these are the books by which we all will one day be judged by the God who made this world, the universe, and all that is. The force within matter that plays strings is the music of the voice of God where the unity of all energy forces is found.

Suggested prayer:

Dear Yahshua Messiah I understand now how Yahuah God spoke everything into existence.
I ask you forgive me of my sins and write my name in the book of life and in the book of remembrance of those who speak and call on your name.
Teach me your ways and lead me by your Spirit.
I surrender my life to you.
I repent of my sins.
I believe in Yahshua Messiah in who's name I pray YahMain.

If you prayed that prayer then you need to do 3 things
1 Find Fellowship in the faith
2 Get Baptized by immersion according to scriptures
3 Read the bible every day

For More information on the true FAITH of the Bible visit our web site and join at

www.YahsSpace.Org

To contact the author write
CYMG
PO BOX 1533
Smethport, PA 16749

GET MORE BOOKS, BIBLES, AND TRANSLATIONS AT

YAHBIBLE.COM

ISBN-13: 978-1519100573

ISBN-10: 1519100574

This book is for educational purposes under laws governing 501 C 3 Non-profit organizations. The use of referenced images for educational materials applies herein.

The Hand Of God In The Time Traveler's Universe; Copyright ©1997-2015 Daniel W. Merrick PHD, Eternal Light & Power Co. Publishing. All Rights Reserved. www.YahsSpace.org

Bibliography:

Bartusiak, Marcia, Thursday's Universe (1988); Davies, Paul, The Cosmic Blueprint (1988); Ferris, Timothy, Coming of Age in the Milky Way (1988); Greenstein, George, The Symbiotic Universe (1988); Harrison, Edward, Darkness at Night (1987); Hawking, Stephen W., A Brief History of Time: From the Big Bang to Black Holes (1988); Judson, H. F., The Search for Solutions (1987); Linde, A. D., Particle Physics and Inflationary Cosmology (1990); Mallove, E. F., The Quickening Universe (1987); Overbye, Dennis, Lonely Hearts of the Cosmos: The Scientific Quest for the Secret of the Universe (1991); Parker, Barry, Creation (1988); Silk, Joseph, The Big Bang (1989); Trefil, James, The Dark Side of the Universe (1988); Abell, G. O., Realm of the Universe (1984); Ferris, T., Coming of Age in the Milky Way (1988); Field, G. B., and Chaisson, E. J., The Invisible Universe (1985); Fredrick, L. W., and Baker, R. H., An Introduction to Astronomy, 9th ed. (1981); Harwit, Martin, Cosmic Discovery (1984); Illingworth, Valerie, The Macmillan Dictionary of Astronomy, 2d ed. (1985); Jastrow, Robert, and Thompson, Malcolm R., Astronomy: Fundamentals and Frontiers, 4th ed. (1984); Lovell, Sir Bernard, and Smith, Sir F. Graham, The Guide to Modern Astronomy (1987); Mitton, Jacqueline, Key Definitions in Astronomy (1982); Pasachoff, Jay M., Astronomy, 2d ed. (1983); Preston, R., First Light (1988); Schaaf, F., The Starry Room (1988); Riordan, Michael, and Schramm, David N., The Shadows of Creation: Dark Matter and the Structure of the Universe (1991); Trefil, J. S., The Moment of Creation (1983); Unsold, Albrecht, and Baschek, R. B., The New Cosmos, 3d ed. (1983); Zeilik, Michael, Astronomy: The Evolving Universe, 4th ed. (1985); Friedlander, M. W., Astronomy (1985); Gribbin, John, In Search of the Big Bang (1986); Wilkinson, D. T., "Anisotropy of the Cosmic Blackbody Radiation," Science, June 20, 1986; Albert Einstein, "Relativity The Special And The General Theory"(1961); Hugh Ross, "The Fingerprint of God" (1991); Beyond 2000, The Discovery Channel, (Feb.1997). Jack Chick Publications. Chick.com MerrickFoundation.Org; FTDNA studies on DNA at FTDNA.com

The Hand Of God In The Time Traveler's Universe; Copyright ©1997-2015 Daniel W. Merrick PHD, Eternal Light & Power Co. Publishing. All Rights Reserved. www.YahsBible.Com

www.ingramcontent.com/pod-product-compliance
Lightning Source LLC
Chambersburg PA
CBHW040851180526
45159CB00001B/390